AA001034

MATERIALS RESEARCH SOCIETY
SYMPOSIUM PROCEEDINGS VOLUME **1602**

Photovoltaic Materials and Devices: Terrestrial and Space Applications

September 16-20, 2013
Kyoto, Japan

Printed from e-media with permission by:

Curran Associates, Inc.
57 Morehouse Lane
Red Hook, NY 12571
www.proceedings.com

ISBN: 978-1-5108-0496-8

Some format issues inherent in the e-media version may also appear in this print version.

©Materials Research Society 2014

This reprint is produced with the permission of the Materials
Research Society and Cambridge University Press.

This publication is in copyright, subject to statutory exception and to the
provisions of relevant collective licensing agreements. No reproduction
of any part may take place without the written permission of Cambridge
University Press.

Cambridge University Press
Cambridge, New York, Melbourne, Madrid, Cape Town,
Singapore, São Paulo, Delhi, Tokyo, Mexico City

Cambridge University Press
32 Avenue of the Americas, New York, NY 10013-2473, USA
www.cambridge.org

Materials Research Society
506 Keystone Drive, Warrendale, PA 15086
www.mrs.org

CODEN: MRSPDH

ISBN: 978-1-5108-0496-8

Cambridge University Press has no responsibility for the persistence or
accuracy of URLs for external or third-part Internet Web sites referred to
in this publication and does not guarantee that any content on such Web sites
is, or will remain, accurate or appropriate.

Additional copies of this publication are available from:

Curran Associates, Inc.
57 Morehouse Lane
Red Hook, NY 12571 USA
Phone: 845-758-0400
Fax: 845-758-2634
Email: curran@proceedings.com
Web: www.proceedings.com

TABLE OF CONTENTS

Fabrication and Characterization of GaAs Tunnel Diode and ErAs Nanoparticles Enhanced GaAs Tunnel Diode for Multijunction Solar Cell..................... 1
T. Sogabe, Y. Shoji, M. Ohba, N. Shunya, N. Miyashita, C.-Y. Hung, A. Ogura, Y. Okada

Mater. Res. Soc. Symp. Proc. Vol. 1602 © 2014 Materials Research Society
DOI: 10.1557/opl.2014.429

Fabrication and Characterization of GaAs Tunnel Diode and ErAs Nanoparticles Enhanced GaAs Tunnel Diode for Multijunction Solar Cell

Tomah Sogabe[1], Yasushi Shoji[1], Mitsuyoshi Ohba[1], Naito Shunya[1], Naoya Miyashita[1], Chao-Yu Hung[1], Akio Ogura[1] and Yoshitaka Okada[1]

[1] Research Center for Advanced Science and Technology (RCAST), The University of Tokyo, 4-6-1 Komaba, Meguro-ku, Tokyo 153-8904, Japan.

ABSTRACT

We report here the fabrication and characterization of GaAs tunnel diode (TD) and ErAs nanoparticles (Nps) enhanced GaAs TD. Four GaAs TDs with different contact area were fabricated by using MOCVD. We found extremely high peak current density of ~250A/cm^2 for the TD with r=0.25mm contact area. Moreover a hysteresis loop was appeared during sweeping up and sweeping down the external voltage. A 'vector load line model' was proposed to explain the origin of the shape of the hysteresis loop and the onset of the bistability occurred at the intersect of the loadline and the current-voltage (I-V) curve of TD. Meanwhile, we have grown ErAs Nps on GaAs(100) surface by using MBE and succeeded in overgrowth of GaAs after ErAs deposition. GaAs(p+)/ErAs(Nps)/GaAs(n+) TDs were fabricated and characterized. We found the GaAs sample containing 70s deposition of ErAs showed the best TD behavior. No TD behavior was observed for the sample without addition of ErAs Nps, clearly indicating the strong tunneling enhancement effect from ErAs Nps.

INTRODUCTION

Among various types of solar cell, III-V multi-junction (MJ) solar cell are the most promising candidate for next generation of high efficiency photovoltaic conversion of sunlight. For instance, $Ga_{0.51}In_{0.49}P/Ga_{0.99}In_{0.01}As/Ge$ triple junction solar cell has reached the recorded efficiency of over 30% under 1 sun and of 40% under concentration ratio of 240 suns [1]. MJ solar cell consists of subcells which are usually interconnected by TDs featured with both low electric resistance and high optic transparence. For MJ solar cell to be operated under light concentration, the enhanced short circuit current density requires the TD to have much lower electric resistance to minimize the voltage drop and much higher peak tunneling current density than the short circuit current of MJ cells [2]. In this work, we report the fabrication and characterization of GaAs TD prepared by MOCVD and ErAs nanoparticles (Nps) enhanced GaAs TD fabricated by MBE. An issue frequently confronted in MBE growth of TD is the insufficient doping concentration. We show that the addition of ErAs Nps into the *pn* junction region is able to compensate this drawback while improving the tunneling probability through the reduction of tunneling distance due to the semimetallic feature of ErAs Nps [3,4].

1

EXPERIMENT

GaAs TDs were fabricated by a multiwafer AIXTRON MOCVD reactor (AIX2800G3) with 8 × 4 in configuration. AsH3, PH3, TMGa, TMIn, and TMAl were used as precursors. In order to investigate the electrode area effect on the tunneling behavior, we prepared four types of TD with circular electrode area ranging from r=0.25mm, 0.5mm, 0.75mm to 1mm. A mesa device structure was employed with both plus and minus electrodes at front side. We have chosen CBr4 for p++ dopant and DMTe for n++ dopant. The doping concentration was confirmed by electrochemical capacitance-voltage (ECV) measurement. For ErAs enhanced GaAs TD, we have grown ErAs Nps on GaAs (100) by MBE at base vacuum pressure of $3 \times 10^{-7} Torr$. The n and p dopants in MBE growth are silicon and beryllium respectively. Erbium is deposited by thermal evaporation and the cell temperature was varied among $1050°C$, $1000°C$ and $950°C$ to adjust the growth rate. During the ErAs deposition, RHEED was used to monitor the epitaxial growth [5-7].

DISCUSSION

Characterization of GaAs TD

Figure 1a show the sketch of the GaAs TD device structure. We fabricated the TD on Ge single junction solar cell prepared on p-type Ge(001) surface with offcut angle of $6°$ from [110]. The dopant concentration and the epilayer thickness were confirmed by ECV showing $1 \times 10^{20} cm^{-3}$ at p^{++} side and $5 \times 10^{18} cm^{-3}$ at n^{++} side (shown in Figure 2b). The thickness of p^{++} and n^{++} layer is 20nm.

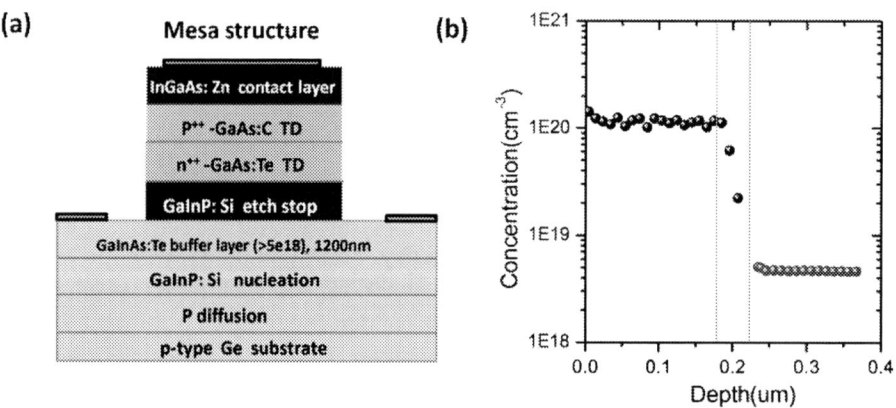

Figure 1: **a**, The device structure of the GaAs TD. **b,** ECV measurement results. The doping concentration for the n region is shown in red and the p region is in black.

I-V curves were measured for the GaAs TDs with four different circular electrodes. As can be clearly seen from Figure 2a, TD with r=0.25mm displayed the highest peak current of $153 A/cm^2$. In Figure 2b, close-up of the low voltage region for the TD with r=0.25mm is presented. If we assume short circuit current of MJ as $15 mA/ cm^2$, we found the voltage drop for concentration ratio X= 1000suns is 0.12V and is 0.05V for X= 400suns. The low voltage drop and high peak current guaranteed normal operation of the TD in MJ cell.

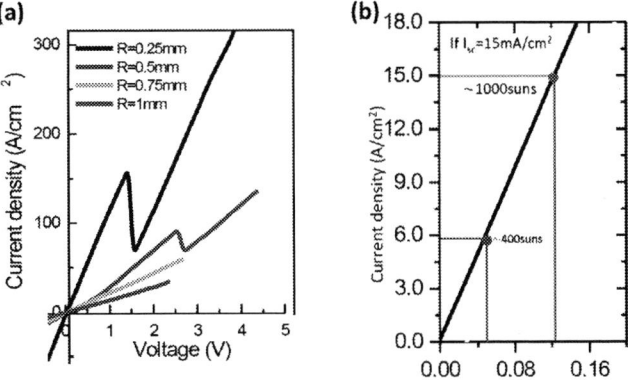

Figure 2: **a**, I-V curves of the four TDs with different contact area of circular electrodes. **b**, Close-up of the TD with r=0.25mm at low voltage

Hysteresis loop in I-V curve of GaAs TD

During measurement of the I-V curve of the TD, a hysteresis loop was open during the voltage sweeping up and voltage sweeping down, as shown in Figure 3. The origin of the hysteresis loop is related to the 'load line' determined by the resistance of the resistor series connected in the electronic circuit [8]. We also found the shape of the hysteresis loop can be tailored by the external resistor artificially inserted into the circuit (see Figure 3a, 3b and 3c).

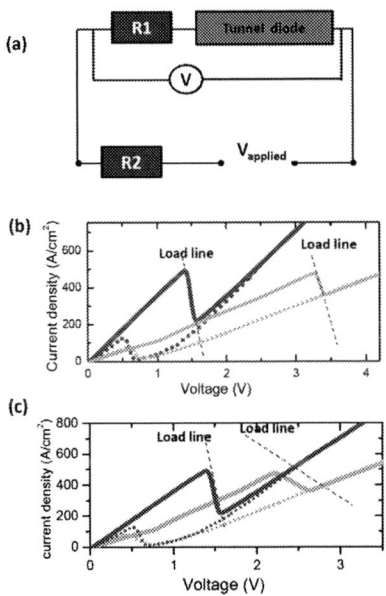

Figure 3: **a**, Electric circuit of the TD measured under different external resistance. **b**, Tailoring the hysteresis loop by adjusting the resistance of R1. R1=0 Ω and R2=0 Ω (purple), R1=3.5 Ω and R2==0 Ω (green). **c**, Tailoring the hysteresis loop by adjusting the resistance of R2. R1=0 Ω and R2=0 Ω (purple), R1=0.5 Ω and R2=3.5 Ω (green)

Deposition of ErAs on GaAs(100) surface

In order to prepare ErAs Nps enhanced GaAs TD, we have at first calibrated the growth rate of ErAs on GaAs (001) surface by using RHEED. Figure 4a shows the (2× 4) RHEED pattern of clean GaAs (100) surface along [-110] azimuth. A three-As-dimer based surface reconstruction model was used to describe the observed (2× 4) RHEED (see Figure 4b). It is known that there exist both 1.6% lattice mismatch and symmetry mismatch between ErAs and GaAs [9]. We have proposed a pseudomorphic model for GaAs overgrowth on ErAs/GaAs (sub) (see Figure 4c).

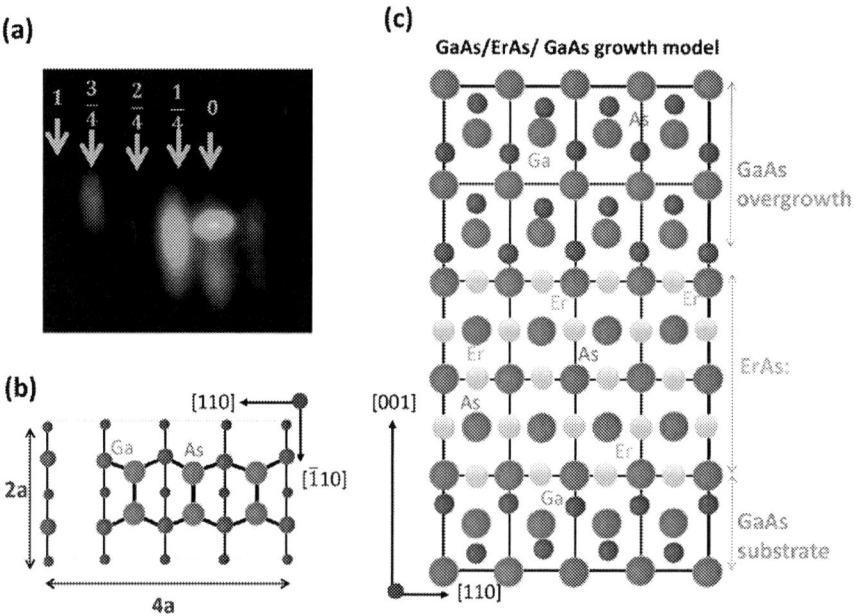

Figure 4: **a,** RHEED of clean GaAs (100) along [-110] azimuth. **b,** Three-As-dimer based surface reconstruction model. **c,** Sketch of a pseudomorphic model for GaAs overgrowth on ErAs deposited on GaAs (100) surface.

In Figure 5, the evolution of RHEED pattern during ErAs growth was monitored. It can be clearly seen that after 200s deposition of ErAs, the RHEED pattern changed from streaky pattern to spotty pattern indicating a growth transition from 2D to 3D. Meanwhile, we also noticed that there existed a recovery of (2× 4) RHEED pattern after 70s of ErAs deposition. This is shown in more detail in Figure 5b, where the streaky line of (× 4) periodicity was disappeared after 20s of ErAs deposition but gradually reappeared after 40s deposition of ErAs and was almost recovered to the initial clean GaAs(001) surface. The recovery of (2× 4) structure validates our proposed pseudomorphic growth model of ErAs on GaAs surface and the sequential epitaxial overgrowth of GaAs on ErAs/GaAs(sub).

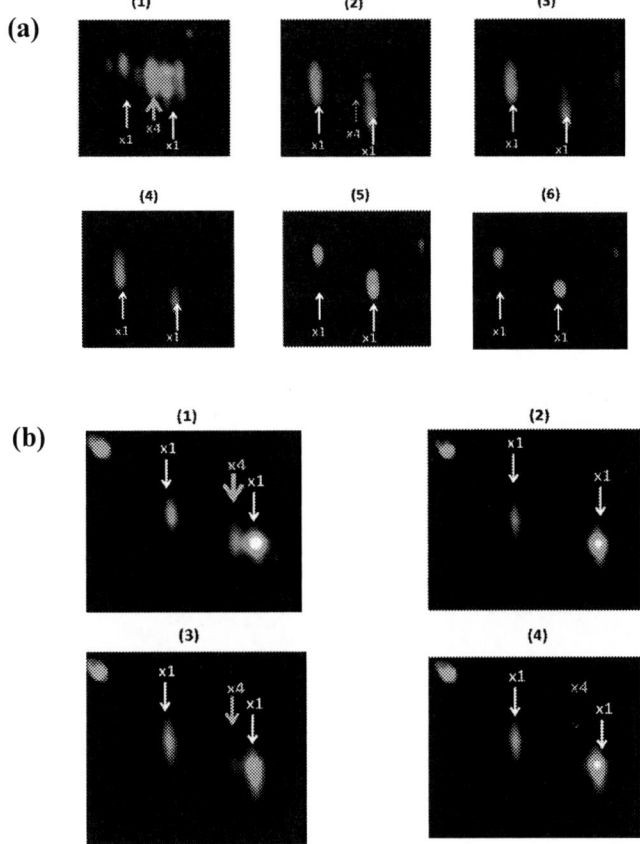

Figure 5: **a,** RHEED pattern recorded during ErAs deposition at the temperature $T_{cell} = 1050\,°C$. Here (1):0s, (2):30s, (3):60s, (4):90s, (5):120s, (6):200s. **b,** RHEED pattern recorded for ErAs deposited at temperature of $T_{cell}=950\,°C$, here, (1):0s, (2):20s, (3):40s, (4):70s.

Characterization of ErAs enhanced GaAs TD

After having calibrated the growth rate of ErAs by RHEED, we prepared two GaAs TD with 70s deposition of ErAs and 200s deposition of ErAs respectively. The TD device structure is shown in Figure 6a where 200nm intrinsic GaAs buffer layer was at first grown on an n-type GaAs(100) surface at substrate temperature of $580\,°C$. Then 200nm of n^{++} GaAs layer was deposited followed by the ErAs deposition and the overgrowth of 200nm of p^{++} GaAs layer at the substrate temperature of $530\,°C$. The I-V curve measurement showed that the GaAs sample without addition of ErAs presented no TD behavior due to the insufficient doping concentration at both the n and p region during MBE growth. In contrast, the GaAs samples with ErAs Nps showed well enhanced tunneling effect. Especially the GaAs sample with 70s deposition

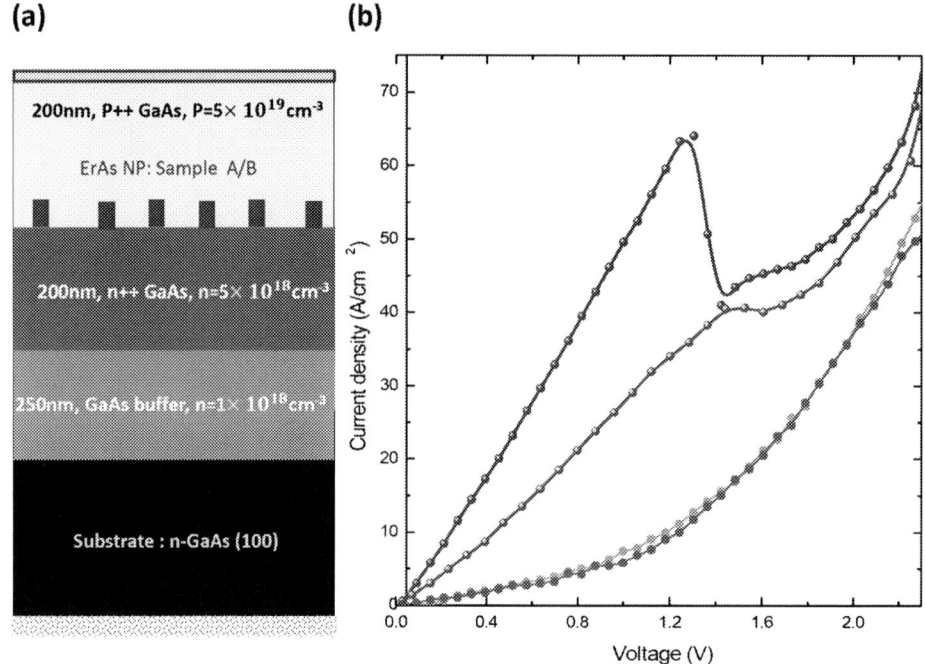

Figure 6: **a,** Device structure of the ErAs enhanced GaAs TD. **b,** I-V measurement results of the two GaAs TD with the addition of ErAs Nps: 70s (blue) and 200s (green). The I-V curves of GaAs without ErAs Nps (red and pink) were also given as reference.

of ErAs displayed better TD behavior than 200s deposition of ErAs while reaching a peak tunneling current density of 65A/cm^2. However, the voltage drop at concentration ratio of 1000suns is about 0.38V and the series resistance is $2.6 \times 10^{-2} \Omega \, cm^2$, which is too large to be used in the MJ cell indicating further improvement is needed to reduce the series resistance of the TD.

CONCLUSIONS

We successfully fabricated and characterized GaAs TD and ErAs nanoparticles (Nps) enhanced GaAs TD. For GaAs TDs fabricated by using MOCVD, we found extremely high peak current density of ~250A/cm^2 for the TD with contact area at r=0.25mm. The voltage drop for concentration ratio X= 1000suns is 0.12V. For ErAs enhanced GaAs TD fabricated by MBE, we found the GaAs TD containing 70s deposition of ErAs showed the best TD behavior. On the other hand, no TD behavior was observed for the GaAs samples without addition of ErAs Nps , clearly indicating the strong tunneling enhancement effect from ErAs Nps.

ACKNOWLEDGMENTS

This work is supported by New Energy and Industrial Technology Development Organization (NEDO), and Ministry of Economy, Trade and Industry (METI), Japan.

REFERENCES

1. R.R. King, D.C. Law, K.M. Edmondson and C.M. Fetzer, G.S. Kinsey, H. Yoon, R.A. Sherif, N.H. Karam, Appl. Phys. Lett. **90**, 183516 (2007).
2. M. Yamaguchi, T. Takamoto and K. Araki, Sol. Energ. Mat. Sol. Cells. **90**, 3068 (2006).
3. A.G. Petukhov, W.R.L. Lambrecht and B. Segall, Phys. Rev. B, **53**, 4324 (1996).
4. H.P. Nair, A.M. Crook, and S. R. Bank, Appl. Phys. Lett. **96**, 222104(2010).
5. C. J. Palmstrøm, N. Tabatabaie, and S. J. Allen, Appl. Phys. Lett. **53**, 2608 (1988).
6. Kris T. Delaney,1 Nicola A. Spaldin,2 and Chris G. Van de Walle2 T. Sands, C.J. Palmstrom, J.P. Harbison, V.G. Keramidas, N. Tabatabaie, T.L. Cheeks, R. Ramesh and Y. Silberberg , Mater. Sci. Rep. **5**, 99 (1990).
7. H.Yamaguchi and Y. Horikoshi, Appl. Phys. Lett. **60**, 2341 (1992).
8. W. Guter and A.W. Bett, IEEE Trans. Electron. Dev. **53**, 2216 (2006).
9. K.T. Delaney, 1 N.A. Spaldin, and C. G. Van de Walle, Phys.Rev.B **81**, 165312 (2010).

NOTES:

NOTES:

Cambridge University Press
32 Avenue of the Americas, New York, NY 10013-2473, USA

Materials Research Society
506 Keystone Drive, Warrendale, PA 15086

ISBN 978-1-5108-0496-8